COLLINS AURA GARDEN HANDBOOKS

GRAPES

D0529043

ALAN TOOGOOD

COLLINS

Products mentioned in this book

Benlate* + 'Activex'	contains	benomyl
'Clean-Up'	contains	tar acids
Fumite General Purpose Smoke	contains	pirimiphos-methyl
'Picket'	contains	permethrin
'Sybol'	contains	pirimiphos-methyl

Products marked thus 'Sybol' are trade marks of Imperial Chemical Industries plc
Benlate* is a registered trade mark of Du Pont's
Read the label before you buy: use pesticides safely.

Editor Maggie Daykin
Designers James Marks, Steve Wilson
Picture research Moira McIlroy

This edition first published 1988 by
William Collins Sons & Co Ltd
London · Glasgow · Sydney
Auckland · Toronto · Johannesburg

© Marshall Cavendish Limited 1985, 1988

British Library Cataloguing in Publication Data

Toogood, Alan R.
Grapes.——(Collins Aura garden handbooks).
1. Viticulture
I. Title
634'.88   SB397.G7

ISBN 0–00–412389–1

Photoset by Bookworm Typesetting
Printed and bound in Hong Kong by Dai Nippon Printing
Company

Front cover: Grape 'Madeleine Angevine'
Back cover: Vitis coignetiae
Both by the Harry Smith Horticultural Photographic Collection

CONTENTS

INTRODUCTION

The cultivation of grape vines, known as viticulture, has been practised for thousands of years.

In the main, grapes are grown out of doors in Europe, in plantations known as vineyards, principally for wine production but also for dessert fruits.

In Britain, also, grapes have been grown outdoors over a very long period, and under glass for well over 200 years, mainly for dessert fruits. More recently, however, there has been an increasing interest in outdoor growing for wine.

The European grape vine (*Vitis vinifera*) grows wild in temperate areas, including Southern Europe, North Africa and Western Asia. Probably a native of South-west Asia, it is a hardy, long-lived, deciduous (it drops its leaves in autumn) climber, capable of growing to a height of 35m (115ft) if unpruned, although hybrids and varieties of it are much shorter.

The plant clings to supports by means of tendrils and bears hand-shaped, lobed leaves 15-20cm (6-8in) across, that take on red and purple tints in the autumn, before they fall.

We do not grow the true species for fruits, but the many named hybrids and varieties upon which fruits ripen in late summer or autumn, depending on variety. Varieties specially suited to growing in cold or heated greenhouses are available, and there are also many for outdoor culture. Outdoor growing is particularly successful in mild areas, but there are varieties which will ripen their fruits even in cooler parts of Britain.

Early grape growing It is thought that wild grapes were used for wine making over 10,000 years ago. Grape growing was well-known to the Ancient Greeks and Egyptians, and the latter trained vines as arches, etc.

The Romans, too, cultivated grapes, growing them on pergolas, as circular tiers, and training them in other forms. They carefully pruned their vines, and propagated them by various methods of layering. By 100 BC they had brought the grape vine to Northern Europe and Britain.

By the Middle Ages, vast vineyards existed in Britain also – often

Two vines grown specifically for their attractive leaf form and spectacular autumn colour.
RIGHT The very popular *Vitis coignetiae*
BELOW *Vitis vinifera* with blue-black fruits.

Vigorous 'Black Hamburgh', arguably the best grape vine for beginners. The large, black fruits have a good flavour and though the vine is at its best when grown in an unheated greenhouse, it is also suitable for a heated one.

around monasteries, until the Dissolution.

Grape vines were grown in heated greenhouses in the 18th and 19th Centuries and were particularly popular in the 19th Century, especially on the big estates where grapes were grown in special greenhouses known as vineries.

In the 1860s a vine pest known as phylloxera devastated the vineyards of Europe, but the European vine was saved from destruction by grafting it onto unaffected American rootstocks. Phylloxera is not found in Britain, fortunately. Outbreaks have occurred in the past, but have been quickly wiped out.

Viticulture today Over the years hybrid grape vines have been produced, resistant not only to phylloxera but the serious disease, mildew. These hybrids are also noted for their heavy crops of fruits, and many named varieties/hybrids are available for outdoor and greenhouse cultivation.

Outdoor growing is becoming increasingly popular in Britain after a period of neglect, as people are finding that grapes can be grown successfully in the open, provided suitable varieties are chosen. Many varieties will ripen their fruits in cool parts of the country, and there is certainly no problem in mild areas with high sunshine records.

There are no problems in growing grapes under glass. Indeed, a great many amateur greenhouses have a grape vine, and this is especially recommended if your part of the country has a particularly unsuitable climate. There are varieties suited to unheated and heated greenhouses or conservatories, artificial heat being supplied, if needed, in spring and when the fruits are ripening in late summer/ autumn. No heat is needed in winter when the vines are resting.

So why not try your hand at growing grapes, whether your interest lies in wine-making, dessert fruits for the table, or ornamental vines (some of which produce edible berries) which you can use to add interest and colour to your flower garden or clothe a sunny wall.

GROWING GRAPES UNDER GLASS

In the main, we grow dessert grapes in greenhouses and conservatories. Some varieties are suited to unheated structures, while others need artificial heat in the spring when growth is commencing, and in the ripening period – late summer and into autumn.

Recommended varieties The following are recommended for growing in greenhouses and conservatories. Details of suppliers are given on page 12.

'Alicante' – ripens very late in the season, and is not widely grown now. But this black grape has a nice flavour, crops heavily. Needs artificial heat to bring out the flavour.

'Black Hamburgh' – highly popular, the best variety for beginners. A black grape, of good flavour, bearing heavy crops on vigorous vines. Reliable in unheated greenhouse, but suitable for heated house. Early to mid-season.

'Buckland Sweetwater' – one of the popular sweetwater varieties. Early, sweet and juicy amber-coloured berries in profusion and of moderate vigour. Ideal for unheated house.

'Chasselas d'Or' – late, maturing mid-autumn. The golden berries, for dessert or wine, have a beautiful flavour. A very popular variety. Needs a heated greenhouse.

'Foster's Seedling' – early to mid-season, a popular sweetwater variety, with heavy crops of amber-coloured fruits, of good flavour, sweet and juicy. Reasonable vigour; suited to cold or heated greenhouses.

'Black Hamburgh'

'Frontignan' – a muscat variety, early, excellent flavoured black fruits. Needs artificial heat.

'Gros Colmar' – late, a black grape bearing heavy crops. Very vigorous and easy to grow. Artificial heat needed to bring out the flavour.

'Lady Downe's Seedling' – late, a black grape of reasonable vigour, but not widely grown now. Artificial heat needed to bring out the flavour.

'Lady Hutt' – late, a sweetwater variety, the white fruits being sweet, juicy and of good flavour; a heavy cropper. Ideal for unheated greenhouses.

'Alicante'

'Lady Hutt'

'**Madresfield Court**' – an early muscat variety, black, very good flavour. Heavy crops on moderately vigorous vines. Needs artificial heat.

'**Mireille**' – early, large white berries, with a muscat flavour. Suited to unheated greenhouse.

'**Mrs Pearson**' – a muscat variety, late maturing, the whitish-yellow fruits having a really good flavour. A strong grower, needing artificial heat, especially during ripening.

'**Mrs Pince**' – a late muscat variety; black fruits of really good flavour; not as widely grown as it used to be. The vines are fairly vigorous. Needs artificial heat.

'**Muscat of Alexandria**' – an extremely popular muscat variety, maturing mid-season. The amber-coloured berries are very sweet and have excellent flavour. Reasonable vigour. Needs warmth during the ripening stage.

'**Muscat of Hamburgh**' – a late muscat variety, heavy crops of red-purple berries with excellent flavour. The vines are reasonably vigorous. Needs artificial heat.

'**Syrian**' – a very late variety with white berries; vines very vigorous. Needs artificial heat to bring out the flavour.

'**Trebbiano**' – a very late variety, with heavy crops of white berries on vigorous vines. Needs artificial heat to bring out the flavour.

'**Mrs Pince**'

'Muscat of Alexandria'

Where to buy vines One can buy the most popular varieties from garden centres, but you will find the selection limited. If you want a wider choice buy from a specialist fruit or vine grower who provides a mail-order service. I recommend:
• Cranmore Vine Nursery, Yarmouth, Isle of Wight. Supply a wide range of greenhouse and outdoor vines for dessert and wine. A descriptive list is available.
• Highfield Nurseries, Whitminster, Gloucester, GL2 7PL. Supply outdoor and greenhouse varieties for dessert and wine. A comprehensive catalogue is available.

Soil preparation One has a choice of growing the vine in a soil border inside the greenhouse or conservatory, or in a border immediately outside the structure, with the main stem (correctly known as a 'rod') growing through a hole in the greenhouse wall, near to ground level.

An inside border means:
• spring growth is earlier because the soil will be warmer, especially if artificial heat is used.
An outside border means:
• less watering, as rain will provide moisture.
• all of the greenhouse floor can be used for other plants.

Vines need a deep, well-drained soil that is slightly acid to neutral, with a pH of 6.5-7.0. Test your soil with one of the soil-testing kits available from garden centres, department stores and chemists selling a range of gardening materials. If below pH 6.5 (acid), add lime.
Dig the border to two depths of the spade – known as double digging – thoroughly breaking the lower soil. Add plenty of well-rotted manure or 'Forest Bark' Ground and Composted.

If your soil is of very poor quality (such as very thin and stony), I would suggest digging out the soil in the border to a depth of about 45cm (18in) and filling up with good-quality topsoil, such as a medium loam, bought from a local supplier. Add approximately half a barrow-load of manure or garden compost to each 1.2m × 60cm (4 × 2ft) area.
If the soil lies very wet or becomes waterlogged in winter, again dig out the soil to at least 60cm (2ft) and put a 15cm (6in) layer of rubble in the bottom to improve drainage. Replace with good-quality topsoil.

If in doubt about the suitability of your soil, check it with one of the amateur soil-testing kits now widely available.

WHERE TO PLANT THE VINE
• In a small greenhouse. Plant at the gable end – the end opposite the door. The rod or main stem can then be grown the length of the house, below the roof ridge.
• In a large greenhouse. Plant on one side, 15-22cm (6-9in) from the wall. The rod can then be trained vertically up the wall and under the roof to the ridge.
• In a lean-to greenhouse or conservatory, plant the vine against the back wall.

When planting a vine, first dig a hole slightly deeper and wider than the container.

Sprinkle in the base of the hole a mixture of fine soil and planting mixture, and dig in, then water.

Remove the container packing, taking care not to damage the roots. Then tease the roots out.

Set the vine in the hole so that the top of its soil is level with the surrounding area.

Work plenty of fine soil in around the spread-out roots (go carefully) and firm well in.

When the 'moat' is filled in completely and the soil is level once more, firm in and water.

How to plant Ideal planting time is late autumn or early winter. Vines are usually pot grown so take out a planting hole slightly deeper and wider than the rootball, add a handful of Growmore, place the plant centrally, work fine soil mixed with an equal volume of 'Forest Bark' Ground and Composted in the space around the rootball and firm it well. If the vine is bare-rooted, then take out a hole sufficiently large so that the roots can be spread to their full extent. Work in plenty of fine soil and 'Forest Bark' Ground and Composted, taking care not to damage the roots, and firm well. Water in thoroughly after planting. If more than one vine is to be grown (say in a large greenhouse or conservatory), space the plants 1.2m (4ft) apart.

How to support a vine Use heavy-duty galvanized wire. Secure the wires horizontally on the wall and under the roof. Space them 22cm (9in) apart and a minimum of 30cm (12 in) from the glass. Any closer and the leaves could be scorched by the sun, and they will not have room to grow properly. Various kinds of brackets can be used to fix wires in both timber and metal-framed greenhouses, and should be available at your local garden centre or, indeed, good hardware shops.

PRUNING GRAPES UNDER GLASS

Vines under glass need annual pruning to prevent a tangled mass of growth and to encourage regular and heavy cropping. We use the 'spur' system of pruning. The plant consists of a single permanent rod or main stem, which can be trained vertically up a wall and into the roof. On this single rod knobs or spurs are encouraged, and these produce growth buds, which grow into shoots, known as lateral or side shoots. The fruits are produced on the laterals, which are cut back almost to the spurs in winter each year. So each year new laterals are produced.

Pruning in year one After late autumn/early winter planting the main stem of the young vine is pruned back fairly hard: the growth which was made in the previous growing season is cut back by two-thirds. If there are any lateral shoots on the remaining stem, cut these back too, to leave one growth bud. When pruning vines, always cut back side shoots immediately above a growth bud.

In the summer of year one the main stem will grow vigorously. Cut out the top when it reaches the top wire. Lateral shoots will also be produced and should be reduced in length by cutting them back to leave five leaves. Side shoots (known as sub-laterals) will be produced from the laterals. These must be reduced to one leaf. The lateral shoots are then tied in to the wires, as is the main stem as it develops.

In late autumn or early winter – as soon as the leaves have dropped – cut back the new growth of the main stem by two-thirds. The lateral shoots should be cut back to leave one growth bud. This is, in effect, the start of spur formation.

Year two From late spring to early autumn the main stem or rod will again grow vigorously; allow it to reach the top wire. You can also allow the plant to produce a couple of bunches of fruits, provided growth is vigorous. Allow two laterals to bear a bunch each. Pinch out the tips of these at two leaves beyond the flower truss. Non-fruiting laterals should be cut back to five leaves, before they become too long. Again tie in as necessary.

In the early winter, when the leaves have fallen, the new growth produced by the main stem is cut back by half. The lateral shoots are cut back to leave two growth buds.

Year three and onwards By the third year, the vine is considered well established. In mid-winter of the third year (and in subsequent years), untie the main stem and allow it to hang down horizontally – the top can almost touch the ground – supporting it with a string attached to the roof. This encourages even breaking of growth buds all along the length of the stem. If the stem is not treated in this way, buds would break near and at the top,

14

leaving the lower part of the stem bare or devoid of lateral shoots. When the buds start into growth re-tie the stem to its normal position.

You can now allow each lateral shoot to carry a bunch of fruits. Prune back the laterals to two leaves beyond a truss of flowers. Allow only one lateral shoot per spur. Leave the strongest shoot and rub out the others at an early stage. Tie in each lateral to a wire, horizontally, but bring each one down to a horizontal position very gradually (over a period of time) or you may snap it out. This means attending to tying several times.

In summer pinch back the sub-laterals to one leaf before they become too long. Any laterals that are not carrying fruits should be cut back to five leaves. When the top wire has been reached cut out the top of the main stem.

After harvesting the fruits (more about this on page 19), reduce the laterals by half their length. Then in winter, after leaf-fall, slightly cut back the top of the main stem or rod, and cut back all laterals to leave one growth bud.

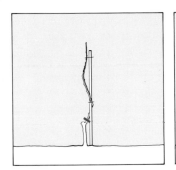

After planting, the main stem of the young vine is pruned hard: the previous season's growth is cut back by two-thirds.

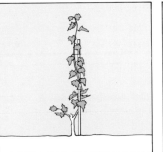

In the summer of Year 1 the main stem grows vigorously. Cut out the top at top wire height; reduce lateral shoots.

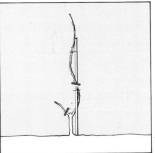

When the leaves have dropped, cut back by two-thirds the new growth of main stem. Reduce laterals to one bud.

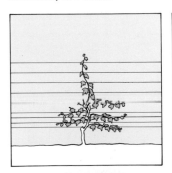

In Year 2 thin out the laterals if necessary. Do this as they develop and you will thus avoid overcrowding.

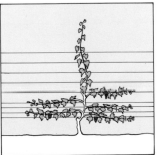

Also in Year 2, a couple of laterals can be allowed to bear fruits: one bunch each. Cut out tips of all laterals.

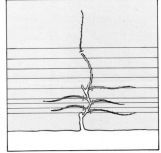

After leaf-fall in Year 2, cut back by half new growth produced by main stem. Cut back laterals to two buds.

Let us now look at the routine care through the growing season (apart from pruning) needed by an established grape vine under glass, commencing with the winter dormancy and ending with the harvesting of the crop of late varieties in the following autumn.

The winter rest The vine needs no heat from the time the leaves fall until early spring, as it will be resting. So keep it completely cold and provide full ventilation, but take care to avoid draughts.

Starting the vine into growth If you are able to provide artificial heat you can start the vine into growth in late winter or early spring. Start with a temperature of about 7°C (45°F) minimum. This can be gradually increased through spring, aiming for a minimum of 18°C (65°F).

If your greenhouse is not heated the vine will come into growth a little later, of course, aided by the warmth of the sun. But by mid-spring the temperature in the house needs to rise and this can be achieved by closing the ventilators. At first, try to aim for a temperature of around 10°C (50°F), with a steady increase to about 18°C (65°F), controlling the temperature by opening and closing the ventilators as and when this is necessary.

ABOVE Lush growth, well-formed fruits.
BELOW Outdoor 'Mueller-Thurgau', here growing under glass.

It is important to keep a vine completely cold during its long winter rest. Make sure the greenhouse is kept well ventilated throughout but not draughty.

Humidity Damp down the floor of the greenhouse when the weather is warm to encourage a humid atmosphere and spray the stems of the vines daily with plain water until flowering starts. However, avoid spraying during periods of strong sunshine to avoid sun scorching young growths. Do not damp down the greenhouse when the vine is in flower as then a dry atmosphere is needed.

You should not have a very humid atmosphere, either, when the fruits are ripening as this can cause splitting. Therefore, it is best to stop damping down at this stage and give some ventilation at night as well as during the day.

Shading From late spring till early autumn provide shade from hot sun. This is most easily achieved by 'painting' the outside of the glass with a proprietary liquid shading material.

Watering When growth commences, heavily soak the border soil with water so that it seeps well down to the roots. Repeat one week later to make sure that the soil is thoroughly moist low down. Then mulch the border with a 5cm (2in) layer of rotted manure, garden compost or 'Forest Bark' Composted.

If the border is inside the greenhouse you will need to water the soil about once a week during warm weather to prevent it from drying out. Always water heavily to soak the soil deeply.

If the roots of the vine are in an outside border, watering may not be needed so often as we can rely on the rain, but certainly apply water generously during dry weather.

When the fruits start to ripen, gradually reduce watering, otherwise the berries may split.

Outdoor variety,
Vitis vinifera 'Brant'   here under glass
in cold area.

Feeding When growth commences, start feeding vines every two weeks, and increase to weekly applications of fertilizer when they are in flower. It is important to stop feeding as the fruits start to change colour – towards the end of the summer or early autumn.

A liquid tomato fertilizer is excellent for vines, as it contains a lot of potash which encourages fruiting. I use ICI Liquid Tomato 'Plus'.

If I find that the vines are making poor growth in summer I apply a fertilizer that contains a high proportion of nitrogen to encourage growth. Alternatively you could use a general-purpose fertilizer: I have found ICI Liquid Growmore to be particularly useful in stimulating growth.

The fruits As soon as vines start flowering a reasonable amount of attention is needed to ensure we end up with well-formed bunches of grapes. It is a fascinating and exciting period in the vine calendar.

Pollination of the flowers To ensure that the flowers set fruits the pollen must be distributed from one to another. This is not as difficult as it sounds, and we do it by hand. The way I do it is to wait until they are fully open and then lightly run my half-closed hand down each flower truss. This distributes the pollen from flower to flower. The atmosphere of the greenhouse should be dry at this period, and that is why I advised you not to damp down the house. Hand pollinate ideally in the warmest part of the day – which is generally around mid-day.

Thinning the fruits Each bunch of fruits will be very crowded with berries, so that there is hardly any room for them to develop. Therefore you also need to remove some of the berries in each bunch.

Early summer is about the time for this thinning out of the fruits, when they are beginning to swell, but before the stalks become hidden from view. There are special vine scissors with long pointed blades available for this job.

The first stage in thinning is to remove most berries from inside the bunch. Then work on the outside, leaving about 5mm (¼in) between the berries to give them room to swell. Further thinning may be

Vine flowers on which the newly forming fruits can clearly be seen. Thinning is not necessary if the grapes are being grown specifically for wine making.

needed later on if you feel that the developing fruits still do not have enough space.

Do not touch the berries when thinning or you will remove the 'bloom' or waxy covering and spoil their appearance. So steady the bunch with a thin forked stick when thinning.

Later in the season, if necessary, cut out with vine scissors any berries which may be beginning to split during the course of ripening.

Harvesting Some varieties (see list of recommended ones) need artificial heat in the autumn when they are ripening their fruits. These are the late varieties and the warmth brings out the flavour. Very late varieties will need artificial heat into early winter. These days, the temperature to maintain must, of course, depend on what you can afford, but do try to keep it around 15.5°C (60°F).

When picking the bunches of grapes again avoid touching the berries if you want the bunches to look really good for the table. Hold them by the main stalk and cut through it with a pair of secateurs.

Keep the bunches in a fairly cool room, and remember that dessert grapes are best eaten soon after picking, when they will be fully ripe and in prime condition.

ABOVE When the flowers are fully open, run your hand very lightly down each truss to set the fruits. RIGHT Thinning out berries from inside the bunch allows for expansion. Vine scissors and a peg-shaped card will assist you to do this without spoiling the 'bloom'.

These days many people are finding that all kinds of fruits can be grown in pots and indeed this is an ideal method for the limited space of today's gardens. The grape vine is an excellent subject for pot culture, and suitably trained, a potted vine is ideal for a small greenhouse. It can be flowered and fruited under glass and then moved outdoors if desired, to make more room under cover for other plants.

I consider it best to use varieties that are suited to growing in an unheated greenhouse, rather than those which need heat in the autumn to encourage ripening. Try the following:

- 'Black Hamburgh'
- 'Buckland Sweetwater'
- 'Foster's Seedling'

Potting Start off with a young vine bought from a garden centre or from a specialist fruit grower (see my recommendations on page 12).

The ideal time to pot the vine is in late autumn, once it is dormant.

Even though plastic pots are widely used today for all kinds of plants, I still consider that clay pots are better for vines, and indeed, for other fruits. Firstly, they are heavier than plastic pots and therefore the plant is less liable to be blown over when it is out of doors.

Second-year growth of potted 'Black Hamburgh'. Its rooting system will be well-established by this stage of its development.

Secondly, there is less risk of the compost remaining too wet in a clay pot, because it dries out more quickly due to the fact that clay pots are porous, allowing air to enter through the sides. It is most important that compost does *not* remain very wet, as this is disliked by vines.

Although there has been a boom in plastic pots over the past few years, many garden centres still stock the clay ones, so you should have no difficulty in obtaining them.

When preparing pots for vines, always place a layer of drainage material in the bottom first. Ideally this should consist of broken clay flower pots (known as crocks). Place a large piece over the drainage hole, and then a 2.5cm (1in) layer of smaller pieces over this. To prevent compost from washing down into the crocks and impeding drainage of surplus water, cover them thinly with 'Forest Bark' Ground and Composted, peat or leafmould.

Choice of compost is important, too. It must be well drained. I use John Innes potting compost No. 3 as this is well drained and quite a rich compost. It is based on loam (soil), peat and sand, and therefore quite heavy, so it supports the plant well and helps to prevent plants blowing over when placed out of doors.

I would advise against the modern soilless or peat-based composts. These are perfectly suitable for many other plants but not for vines and other fruits in pots, because they are very lightweight and would not support the rather heavy plant. Also, you cannot firm them too much, otherwise they hold too much water and remain very wet. In some, the plant foods are quickly used up, so one would have to start feeding sooner, and more regularly.

Do not put a young vine immediately into a large pot, but rather pot on gradually. If the vine has a large volume of compost around its roots, it will remain too wet.

So start off a young vine in a 15cm (6in) pot. First place a layer of compost in the bottom over the drainage material and firm it well. Remove the plant from its nursery pot and place it centrally in the new pot. Trickle compost around the rootball, firming well as you proceed. Leave space at the top of the pot for watering in after potting.

In the second year the vine will be well rooted in this pot, so can now be moved to a 20cm (8in) pot. Finally the vine can go into a 30cm (12in) pot. Again, this should be done in late autumn.

When potting vines, ensure good drainage by using a generous layer of crocks, covered by well-firmed compost. Position the vine centrally and add more compost, firming as you go.

TRAINING AND PRUNING POTTED VINES

Vines in pots can be trained to various shapes – indeed the vine is most adaptable in this respect. However, I always use the spur system of pruning: an adaptation of the system described earlier on page 14. Basically, there is a permanent framework on which spurs develop, and produce side or lateral shoots to carry the bunches of grapes.

In the second year allow each pot-grown vine to produce two bunches of fruits, the rest being removed at an early stage. Young vines should never be allowed to become over-loaded with fruits.

In subsequent years, allow one bunch of fruits per lateral. The spurs are spaced 30cm (12in) apart on a pot-grown vine, again so as not to overload it with fruit. An average pot-grown vine is capable of bearing five to eight bunches of grapes. Therefore, you may need to grow several plants if you want a really good supply of grapes. Even so, they take up little space.

Spiral fashion A very popular method is to train the main stem spiral fashion around a framework of stout bamboo canes. One can go to a height of about 2m (6ft). I would not recommend going any higher, as then the plant can become rather unmanageable when you need to move it around.

Side shoots should be thinned out to 30cm (12in) apart. Try to ensure they are evenly spaced. The side shoots are cut back to two leaves beyond a bunch of fruits. In the winter the main stem or rod is reduced by about half its length and the side shoots cut back to one or two buds.

Standards Alternatively, a grape vine in a pot may be trained as a standard – rather like a standard rose, fuchsia or pelargonium. Basically this consists of a straight permanent stem, free from side shoots. The shoots are allowed to grow at the top of the stem to form a 'head'. The height of the stem should be around 1.5m (5ft). Allow a system of spurs to form at the top, and again prune lateral shoots to within one or two buds of these in winter.

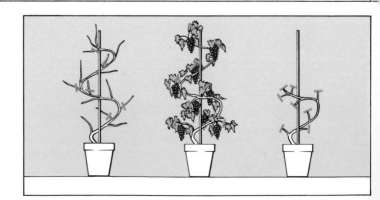

Lack of space need not preclude you from growing a vine. All you actually need is a pot, a frame of strong bamboo canes and good compost. Here you can see the spiral fashion pruning outlined in text above.

Allow about five or six lateral shoots to bear fruits each year. The laterals will need tying to the bamboo cane that supports the stem of the plant, to prevent them from snapping out. Tie them in before the flower trusses are too far developed.

General care The pots can be stood out of doors and taken into the greenhouse in late winter to start the vines into growth. If possible, provide some artificial heat. When the fruits have been harvested the vines can be placed out of doors again for their winter rest.

If you do not have a greenhouse, grow the vines in a sunny, sheltered spot out of doors. In this instance choose a variety suitable for outdoor cultivation. There's a good selection available, as you will see on pages 24-6.

All the other aspects of cultivation such as watering, feeding, temperatures, humidity, etc. are the same as for permanent greenhouse vines, with one exception – that of top-dressing. In the spring each year carefully scrape off about 2.5cm (1in) of compost and replace with an equal amount of fresh compost.

Third-year pot cultivation of a grape vine. Here, 'Golden Chasselas', bearing well-spaced bunches of thinned-out fruits.

GROWING GRAPES OUT OF DOORS

Commercial vineyards are becoming a familiar sight in Britain, and so, too, are smaller versions in private gardens, for it is certainly not true that these islands are too cold and have insufficient sun to ripen grapes. You can grow grapes for wine and/or dessert, if you choose varieties suited to your part of the country.

The grape vine is hardy in all parts of Britain but likes a warm sheltered aspect. In some of the colder northern areas, protection is recommended to ripen the fruits. Viticulture is very successful in the mild south, south-west, the west coast and in mild areas of Ireland.

The black varieties need most warmth and a long season, but less warmth is needed by the white varieties, so these are an ideal choice for cooler parts of Britain. Indeed, most of the grapes which are grown out of doors here are white varieties.

RECOMMENDED WHITE GRAPES

'Chardonnay' – a good wine grape, ideal for cool areas. Grows well in East Anglian vineyards.

'Himrod' – a seedless grape with golden berries. Needs a sunny, warm wall for success.

'Madeleine Angevine 7972' – grape maturing early autumn, for dessert or wine. Reasonable flavour and gives heavy crops of pale green fruits. Vigorous, but gets mildew. Good for cooler parts of the country.

'Madeleine Angevine'

A cooler, more difficult area; grapes best grown under glass

Ideal grape-growing areas

Cooler areas but still suitable for grapes, particularly if white varieties are grown. May need protection to ripen fruits in the coldest of these areas

'Madeleine Sylvaner 2851' – very early, an excellent variety for dessert or wine. Crops well. Ideal for cooler areas.

'Mueller-Thurgau' – matures mid-autumn. Good for wine, heavy crops even when young; vigorous, golden-brown berries. Very widely planted and highly recommended; a reliable cropper even in cool areas.

'Mueller-Thurgau'

'Muscat de Saumur' – an old French grape grown for dessert, with a very fine muscat flavour. Needs a warm wall.

'Pinot Blanc' – late maturing. Considered to be an excellent wine grape. Good for cool summers.

'Précoce de Malingre' – early autumn. For dessert or wine, nice flavour. Small crops, moderate vigour. Needs warm, sheltered position.

'Seyval' – matures mid-autumn. For wine, good flavour. Crops heavily, moderate vigour. A warm, sheltered position is required.

'Seyve-Villard 5/276' – early, golden berries for both dessert and wine.

Vigorous, heavy cropper. Likes limy soil, is popular and a good companion for 'Mueller-Thurgau'. Needs a warm, sheltered position.

'Seyve-Villard'

'Siegerrebe' – matures late summer/early autumn. Golden berries with nice flavour, for wine; but excellent too, for dessert. Crops heavily, growth moderately vigorous. Not recommended for chalky soils.

'Siegerrebe'

'Traminer' – late in a cool summer. Grow on a warm sunny wall or fence; its rosy-coloured berries make a spicy, full-bodied wine.

RECOMMENDED BLACK GRAPES

'Baco 1' – a good wine grape. Very vigorous, heavy cropper. Try it on a large wall.

'Brant' – matures mid-autumn. Good flavour for wine. Heavy crops, vigorous. Excellent for a warm wall.

'Leon Millot' – an excellent wine grape. Heavy cropper, very vigorous. Try it on a large wall.

'Marshal Joffre' – similar to 'Leon Millot', but earlier.

'Pirovano 14' – early, fine flavour. Ideal for dessert or wine. Needs warm, sheltered site.

'Seibel 13053' – early, heavy cropper, strong grower. Highly recommended for red/rosé wines. Warm, sheltered site required.

'Schuyler' – an excellent dessert grape of good flavour. Vigorous; best on a warm wall or fence.

'Strawberry Grape' – matures mid-autumn. A dessert grape with a reasonable musky flavour, reasonably heavy crops, moderate vigour. Grow against a warm wall.

'Triomphe d'Alsace' – early, vigorous and disease-free heavy cropper. A very successful red-wine producer on a warm, sheltered site.

'Seibel'

26

Shelter and aspect Outdoor grapes need to be grown in full sun, and not shaded by trees, buildings, etc. Make sure the site is sheltered from the wind. If this is a problem erect a windbreak on the windward side. This could be a screen of windbreak netting supported on posts about 2.4m (8ft) high, or it could be a living windbreak, such as a screen of conifers like the Leyland cypress (*X Cupressocyparis leylandii*).

● Elevation ideally should not be over 91m (300ft).

● If available, a south- or west-facing slope is ideal, with the rows running from north to south.

● Avoid planting vines in frost pockets – hollows and low-lying areas, such as valley bottoms. Cold air drains down into these and collects at the bottom, resulting in very frosty conditions.

● In very cool areas the vines would be better grown against a south-facing wall or fence.

Suitable soils Grape vines can be grown in a wide range of soils, provided they are deep, fertile and very well drained. If drainage is a problem you will need to lay a herring-bone system of clay land-drain pipes, emptying into a soak-away at the lowest point. Modern, plastic land-drain pipes are available to make the job easier.

You will also need to test your soil for acidity or alkalinity before planting, and adjust it if necessary. A recommended pH for vines is 6.5-7.0, which means the soil is slightly acid to neutral. In other words, it contains little chalk or lime. Not many vines relish a limy or chalky soil. A chalky soil would result in the vines suffering from chlorosis, when the leaves turn yellow and growth is generally poor and stunted. If the pH is lower than 6.5 you will need to apply lime to raise it. Soil can easily be tested with one of the inexpensive kits available at garden centres.

'Schuyler'

Preparing the soil About two months before you plant a vine, the soil should be dug deeply so that the vines have a deep root run. Therefore dig the site to two depths of the spade (this is known as double digging), and thoroughly break up the lower soil, which may be hard and compacted. At the same time remove all perennial weeds (including the roots). If these are a real problem, it would be advisable to treat them with glyphosate weedkiller when they are in full growth, and several months before commencing digging.

Either 'Forest Bark' Ground and Composted or rotted manure should be added to each trench during digging. After digging, apply lime if necessary (according to the pH test) and work it into the surface.

A week or so before planting apply a general-purpose fertilizer such as Growmore according to maker's instructions and lightly fork this into the soil surface. The land is now ready to receive the young vines.

Providing supports Outdoor vines need adequate supports but the system will depend on how they are being grown.
● Vines on a wall or fence. Fix to the wall or fence horizontal wires spaced 30cm (12in) apart, and at least 3.8cm (1½in) away from the wall. Use heavy-duty galvanized wire, or plastic-coated wire. To fix wires to a wall use vine eyes or masonry nails, inserted every 1.5-1.8m (5-6ft). For fences one can use the screw-type vine eyes. All of these are available from good garden centres.
● Vines in the open ground. Here the vines are grown in straight rows and need a system of posts and wires. Use strong fencing posts, first

ABOVE Outdoor vines fixed to well-spaced wires on a sunny wall.
RIGHT Vines grown on the Guyot system – the most widely-used method of training.

treated with a good, non-creosote timber preservative such as Cuprinol. Use 1.9m (6½ft) long posts and insert them 60cm (24in) deep in the soil, to give you a height of 1.3m (4½ft). Space the posts 2.4m (8ft) apart. The end posts should be made secure with angled timber struts.

The posts support heavy duty galvanized wires spaced 30cm (12in) apart, the lowest wire being 30cm above ground. The wires need to be really tight, and this is best achieved by using straining bolts at each end. The wires can be stapled to the intermediate posts.

Each plant will need a stout bamboo cane, long enough to reach the top wire after being pushed well into the ground. Tie the canes to the wires with strong twine.

Planting Buy one-year-old vines and plant between mid-autumn and late winter. If they are in containers all you need do is take out a hole slightly wider than the rootball, place the plant in the centre, and so that the top of the soil is just 12mm (½in) below ground soil level. Work fine soil between the rootball and the sides of the hole, and firm it well.

If the vines are bare rooted take out a hole of a sufficient width to allow the roots to be spread out to their full extent and to the depth indicated by original soil line at base of vine. Work fine soil between them and firm well with your heels.

If the vines have been grafted (which is not often the case these days) make sure the graft union (the swollen area near the base of the stem) is above soil level.

After planting tie in the stem to the cane.

Spacing of plants

● If the vines are being grown against a wall or fence space them 1.2m (4ft) apart and 20cm (8in) away from the wall.

● Vines in the open ground are spaced 1.2-1.5m (4-5ft) apart in the row. Rows should be spaced 1.5-1.8m (5-6ft) apart.

If the soil is dry, water in the newly planted vines. Then mulch the soil around them with a 5cm (2in) layer of well-rotted farmyard manure or garden compost, to prevent the soil from drying out rapidly and to suppress the growth of annual weeds.

Double digging entails making a trench 90cm (3ft) wide by 25cm (10in) deep, and required length.

Break up the bottom 25cm (10in) of soil with a fork and add a generous layer of manure.

Double dig the adjoining strip, throwing the soil sideways into the first trench. Repeat to end.

PRUNING GRAPES OUT OF DOORS

Outdoor vines must be pruned regularly, otherwise they will become a tangled mass of growth. Furthermore, unpruned vines will start to deteriorate in vigour and therefore crops will be lighter. The method of pruning depends on whether the vines are grown on walls/fences, or in rows. Let us first consider the spur system for wall-trained vines.

We use the spur system of pruning for vines grown on walls and fences, whereby we have a permanent stem or rod which produces lateral shoots each year from woody spurs or knobs.

On a wall or fence there may well be more room and height for the vine than when grown under glass. So there is no reason why we should not let the rod grow to quite a height if space permits, say on the wall of the dwelling house.

Early pruning Immediately after planting, the main stem is cut back hard. Two-thirds of the wood that was produced the previous summer is removed, and any side shoots are cut back to one bud.

In the first growing season the main stem will grow quite vigorously – maybe to 2.4m (8ft) or more, and it will also produce side or lateral shoots. These side shoots should be cut back in the summer to five leaves, otherwise they will result in a tangled mass of growth. Side shoots or sub-laterals which are produced from these should be shortened to one leaf.

As the main stem and laterals develop tie them in to the wires, using soft garden string.

ABOVE 'Siegerrebe', a moderately vigorous vine, shown here being pruned.
RIGHT The same vine after pruning: the main stem and two laterals remain.

RIGHT 'Siegerrebe', here being grown on a large scale, before pruning begins.

When the leaves have fallen in late autumn or early winter cut back the new growth of the main stem by two-thirds of its length and shorten the laterals to one bud.

Year two The main stem will again make good growth and laterals will develop. Allow two or three of the laterals to bear fruits. Pinch them out at two leaves beyond the flower truss. Other laterals should be cut back to five leaves.

Then in late autumn or early winter cut back the new growth on the main stem by half and cut back the laterals, leaving just two buds.

Year three and onwards In the third year and onwards normal pruning is carried out, as the vine is well established.

Allow one side shoot to develop on each spur. Pinch out the laterals at two leaves beyond a flower truss and carefully tie them in to the wires. Sub-laterals are pinched back to one leaf. The main stem should be stopped (have its top cut out) at the desired height. In late autumn or early winter prune back all lateral shoots to one bud.

GUYOT TRAINING AND PRUNING

The Guyot system is used for vines grown in rows in the open ground as opposed to cultivation against a wall or fence. Actually there are several variations on this system but undoubtedly the most popular is the double Guyot, so this is the technique described here. It is called the double Guyot because each plant has two main branches.

Basically, we let three main stems develop on each plant every year. We retain two of these, which bear the fruits. The third stem is pruned back, when it will produce the replacement fruiting stems for the following year. The fruiting stems are trained near the ground.

Initial pruning and training After planting cut back the main stem to within 15cm (6in) of the ground – or 15cm (6in) above the graft union in the case of grafted plants. It is essential to leave two growth buds.

From spring to early autumn, growth will be made. During this period the strongest shoot is trained up the cane. Other shoots are cut back to one leaf.

In late autumn the stem is cut down to 38cm (15in). It is essential to leave three growth buds.

Second year pruning In the summer three shoots are trained vertically, tying them in to the cane. Lateral or side shoots are cut back to one leaf.

In late autumn commence pruning as for an established vine. Tie down one shoot horizontally to the

LEFT Guyot-trained vines as they should appear during the dormant period.
ABOVE Guyot-trained vines being grown on a large scale in a commercial vineyard. Net being used here as a protective measure obscures the form, but this can be seen clearly at left.

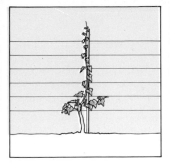

Spring growth from winter-planting. Cut back the side shoots. Train the central one vertically.

In November cut back the central rod to 38cm (15in) and shorten the lateral growths to just one bud.

Year 2. Train the three shoots vertically, pinching back the new laterals to within one leaf of the stem.

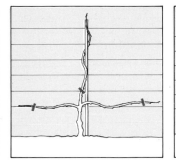

In November, cut back the top shoot and ease the two side shoots outwards to wire; tie them in.

Year 3. Fruit bearing laterals are always trained vertically. Stop fruiting laterals when they grow above the top wire.

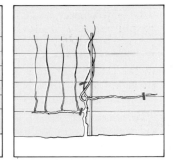

In late autumn remove the lateral bearing rods. Cut back central vertical and tie in the others to the wire.

left and another to the right, securing them to the lowest wire. They should then be cut back to leave at least 60cm (2ft) of growth. The third shoot is cut down to leave three growth buds. These buds will produce the replacement stems for the following year.

Year three and onwards During the spring to late summer train three shoots from the centre up the cane (vertically), and if they produce laterals pinch these back to 2.5cm (1in). The fruit-bearing lateral shoots which form on the two horizontal stems are trained vertically – tie them in to the wires. When they grow above the top wire cut them back to three leaves above the top wire and cut out any sub-laterals so that only strong laterals are left.

In late autumn cut out the old fruited stems, back to the replacements. Tie in the replacements to the lowest wire – one to the left and one to the right. Cut back the soft tips of each to leave at least 60cm (2ft) of growth. Cut back the remaining stem to leave three growth buds.

CARE OF GRAPES OUT OF DOORS

Outdoor grape vines do not require a great deal of attention apart from pruning but there is no doubt that they greatly benefit from regular feeding and watering during the growth season. A regular food supply encourages not only good vegetative growth but assists in the production and ripening of fruits. A steady water supply will ensure the berries develop properly and are juicy when ripe.

Feeding Apply a general-purpose fertilizer to the soil in late winter and lightly prick it into the surface. This will supply the major plant foods – nitrogen, phosphorus and potash. I use Rose 'Plus', according to the manufacturer's instructions on the pack.

Vines need a good supply of potash, for it is this food that encourages fruit production and ripening, so I also apply sulphate of potash in late winter at 14g per sq m (½oz per sq yd) and lightly prick it into the soil surface.

In the spring each year mulch the soil around the vines with bulky organic matter, for example 'Forest Bark' Ground and Composted. A layer about 5cm (2in) deep will help to prevent the soil drying out rapidly and suppress the growth of annual weeds. A mulch should always be applied to moist, weed-free soil, so hoe off any weeds.

Also in the spring, when the vines are in leaf, spray them with a solution of Epsom salts (magnesium sulphate), to prevent magnesium deficiency. If vines suffer from a deficiency of this nutrient, reddish bands appear between the leaf veins and growth is poor.

Make up a solution of 226g Epsom salts in 11 litres of water (½ lb in 2½ gallons). Apply it to the vines ideally with a garden pressure sprayer, or with a watering can

Spring spraying of a vine, with a solution of Epsom salts, to prevent magnesium deficiency.

fitted with a fine rose. Thoroughly wet the foliage. Repeat the application a fortnight later.

In the growing season I find that dessert grapes benefit from weekly liquid feeding applied to the soil around the plants until the berries start to ripen. I use ICI Liquid Growmore, according to the manufacturer's instructions on the bottle.

Watering Never allow the soil to dry out, because this can seriously affect fruit production. It is particularly important to keep a close eye on vines grown against a wall, as the soil can dry out very rapidly immediately in front of a warm sunny wall.

At each watering give a really thorough soaking, so that water penetrates deeply into the soil, moistening it to a depth of at least 15cm (6in). This means applying the equivalent of 2.5cm (1in) of rain – say 27 litres of water per sq m (4¾ gallons per sq yd). The best way to apply such a large amount is to use a garden sprinkler attached to a hosepipe. You can check how deeply the water has penetrated by digging a hole with a hand trowel, about 15cm (6in) deep. If the soil is still dry at the bottom then apply more water. Carry out this test about an hour after the initial watering.

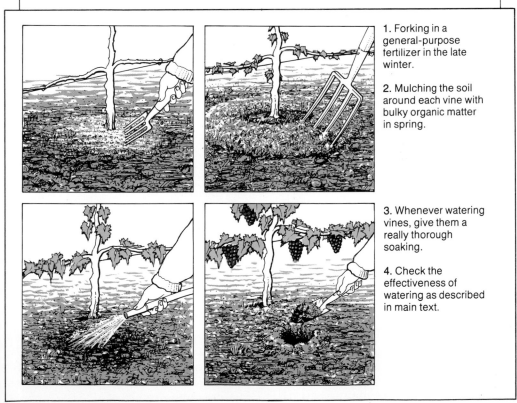

1. Forking in a general-purpose fertilizer in the late winter.

2. Mulching the soil around each vine with bulky organic matter in spring.

3. Whenever watering vines, give them a really thorough soaking.

4. Check the effectiveness of watering as described in main text.

Removing surplus fruit bunches
If the vines are vigorous and healthy you will find that many bunches of fruits will be produced on each plant. It is not advisable to allow all of these to remain, as they will be a strain on the plants and result in poor-quality fruits. If a vine is overloaded in one year you will probably find that in subsequent years it produces few if any fruits.

Therefore, some bunches should be removed at an early stage of their development. Obviously one should leave those bunches which are of good shape, of adequate size, and with plenty of berries in the bunch. Poorly shaped, thin bunches are the ones to cut out.

With the spur system of pruning one should allow only one bunch of fruits to develop on each lateral shoot.

Do not allow young vines to carry too many bunches. A three-year-old vine should be allowed to carry no more than two or three bunches. A four-year-old plant can carry four or five bunches. Then in subsequent years allow full or normal cropping.

THINNING THE FRUITS
- This is not needed if the grapes are being grown for wine.
- Dessert grapes will need thinning, as described on pages 18-19 to ensure good-sized fruits.

Removing leaves In the early autumn it is a good idea to remove the lower leaves of the plants gradually, over a period of several weeks. Leaf removal ensures that the bunches of grapes receive plenty of sun to assist in ripening.

Leaf reduction also improves the circulation of air around the fruits, thus preventing trouble from the fungal disease grey mould (botrytis), which can cause bunches of fruits to rot, and spreads rapidly by contact.

Damaged fruits Inspect the bunches of fruits several times a week, for berries can become diseased, split, or damaged by birds. If any such damage is noticed cut out the affected berries with a pair of vine scissors as shown above.

ABOVE When you are cutting out bunches, handle carefully to avoid spoiling bloom.

LEFT Thinning out a dessert grape with the aid of vine scissors.

ABOVE Pinching out unwanted young shoots on the laterals.

BELOW Removal of excessive foliage allows a vigorous vine to fruit well.

Harvesting grapes It is very tempting to pick the fruits immediately they have their full colour, but this should be resisted because they will not yet contain sufficient sugars. Grapes need four to five weeks for the sugars to form – that is, for the early and mid-season varieties. Late-maturing varieties need even longer – eight to ten weeks.

If the grapes are being grown for dessert it is important to avoid handling or touching the actual berries for this rubs off the natural bloom – a whitish, waxy coating. If this is accidentally rubbed off it will spoil the appearance of the bunch.

Each bunch should be cut off the vine with a short piece of stem attached to the stalk. This effectively forms a handle, enabling you to hold the bunch of fruits without touching the berries.

ORNAMENTAL GRAPE VINES

Some grape vines are very decorative and are grown in the ornamental garden. Again, they are climbing plants, gripping their supports with tendrils, and the leaves often take on rich autumn colours.

The ornamental vines are very effective when allowed to grow up a large tree or through an old hedge. Also, gardeners often use an ornamental vine for covering an unsightly tree stump.

Use them, also, for covering the walls of the house or garage and for fences and garden walls. Try growing ornamental vines, too, over low walls, or use them as ground cover to clothe a steep sunny bank.

I like to grow the ornamental vines with large-leaved ivies, as the two contrast beautifully, especially if allowed to intertwine. Try a vine with the variegated Canary Island ivy (*Hedera canariensis* 'Gloire de Marengo'), with deep green, silvery-grey and creamy-white evergreen leaves; or with the variegated Persian ivy (*Hedera colchica* 'Dentata Variegata'), whose large evergreen leaves are edged creamy-yellow to creamy-white. A superb effect is created when the vines take on their autumn leaf colour.

Although all of them produce grapes, not all are recommended for use, but fruits of the variety 'Brant' make good wine and can also be used as dessert grapes.

ABOVE *Vitis coignetiae* as the foliage takes on its stunning autumnal hues.

LEFT *Vitis* 'Brant' trained as a path-side shrub in a sunny front garden.

SPECIES AND VARIETIES

Vitis amurensis – the Amurland grape. This is a vigorous species whose young shoots are reddish and very attractive. The large-lobed leaves turn crimson and purple in the autumn, contrasting with the black fruits. Not suitable for eating.

Vitis betulifolia – the lobed leaves of this species take on rich autumn colours and blue-black fruits are produced. Not suitable for eating.

Vitis 'Brant' – the most popular ornamental vine, a tall and vigorous grower, is cultivated mainly for its sweet purple-black fruits, but the crimson autumn leaf tints are quite attractive.

Vitis coignetiae – this species bears huge leaves up to 30cm (12in) across, and they have the most brilliant autumn colour: crimson and scarlet. A magnificent effect can be created when this vine is grown with large-leaved variegated ivies. Large black fruits are produced, covered with purple bloom, but not suitable for eating. This is undoubtedly the most spectacular of the ornamental vines – hence its popularity.

Vitis amurensis

39

Varieties of the ornamental vine *Vitis vinifera* bear edible fruits as a bonus to their decorative foliage.

Vitis davidii – a vigorous species with spiny shoots, and heart-shaped leaves that are deep green and glossy above, greyish green and bristly below. They produce excellent autumn colour – rich crimson. Fruits not suitable for eating.

Vitis pulchra – the shoots of this species are reddish and the toothed leaves are also reddish when they first unfold. In the autumn the leaves turn brilliant scarlet. Fruits not suitable for eating.

Vitis vinifera 'Incana' – this is known as the Dusty Miller grape because the grey-green leaves are covered in white down. The black fruits are edible.

Vitis vinifera 'Purpurea' – this is popularly known as the Teinturier grape. Its leaves are the most beautiful shade of claret-red when they are young, and as they age they turn deep purple. A good companion plant is a red climbing rose. The grapes are edible.

Suppliers
Hillier Nurseries (Winchester) Ltd, Ampfield House, Ampfield, Nr. Romsey,
Hampshire, S05 9PA.

Notcutts Nurseries Ltd, Woodbridge,
Suffolk, IP12 4AF.

GENERAL CARE OF ORNAMENTALS

The ornamental grape vines need the same conditions as other outdoor vines, as described on pages 24 and 26. Certainly a warm sunny spot is required, these vines being ideal subjects for sunny south- or west-facing walls and fences. Good drainage and well-prepared soil are essential.

Supports will be required if growing on a wall or fence, so erect horizontal galvanized wires spaced 30cm (12in) apart. Or you may prefer to fix trellis panels to walls and fences and train the vines to these. One can obtain trellis panels in timber, or plastic-coated steel mesh. Vines are also excellent subjects for free-standing trellis screens, pergolas and arches.

Feed and water the ornamentals as described under outdoor fruiting vines, see pages 34-5. And make sure they have good drainage.

Pruning If ornamental vines have unrestricted space – for example, if they are being grown up a large tree or against a high wall – they need little attention after the post-planting pruning.

Immediately after planting reduce the new growth of the main stem by about two-thirds of its length.

If the vines are being grown in restricted space they will need annual winter pruning to keep them within bounds. The new growth is pruned back to a permanent framework of older wood or stems, trained to fill the available space. There is no need to stick to a single main stem with the ornamentals.

Carry out pruning in the early winter – not when growth is commencing or the pruning cuts may 'bleed' – lose sap – profusely. This, of course, has the effect of weakening the plants.

The young wood produced in summer is cut back to one or two growth buds near its base. This type of pruning results in woody spurs building up on the main framework – these spurs are attractive in their own right and of course show up particularly well in the winter when the leaves have fallen.

You will find that the bark, often a warm reddish brown, starts flaking or peeling off in strips as the main framework ages. This is perfectly natural, and looks most attractive, showing to best advantage in winter. So even in the dormant season the ornamental vines add interest to the garden.

If the vine is in a restricted space, you may need to carry out summer pruning, too, as the side shoots are often very vigorous and are capable of growing quite long in a single season.

So, if necessary, prune in mid-summer, cutting back the shoots to five or six leaves. Sub-laterals may then be produced and if these become too long pinch or cut these back, too. Then winter pruning follows as outlined above.

Pruning tools Throughout this book there is a lot of information on pruning, so here is advice on the most suitable tool to use.

All you need is a pair of really sharp secateurs. Buy good-quality ones as then the blades will remain sharp for a long time.

There are two types of secateur – the parrot-bill and anvil. Use parrot-bill secateurs for pruning vines as they make cleaner cuts.

Grape vines are propagated most usually from cuttings known as 'vine eyes', a suitable method of propagation for fruiting and ornamental vines, best carried out in early to mid-winter. But there are other methods: from long hardwood cuttings and by layering.

A vine eye consists of a short piece of stem with a dormant growth bud or 'eye', rooted in heat. A good percentage take is achieved in a temperature of 21°C (70°F); use an electrically heated propagating case if you can.

Use well-ripened hard stems produced in the previous growing season, and cut them up with sharp secateurs. Each piece should be 2.5-3.5cm (1-1½in) long. Make the top cut just above a bud and the bottom cut between buds. Leave only one bud at the top of the cutting – carefully cut out the opposite bud.

Alternatively, each 3.5cm section can have a bud in the centre. In this instance, remove a sliver of wood on the side opposite the bud, to ensure better rooting.

Dip all the cuttings in 'Keriroot' hormone rooting powder, then root them in a mixture of equal parts peat and sharp sand. Cuttings with a bud at the top should be inserted vertically, so that the bud rests at compost level. Cuttings with a bud in the centre are inserted horizontally, so that the bud faces upwards and is just above the level of the compost. Water the cuttings well in and place in a propagating case.

Hardwood cuttings These are taken at the same time of year as vine eyes, using the same type of material and providing the same conditions of rooting.

LEFT Young vine that has been raised from an 'eye'.

ABOVE Newly-planted vine 'eyes' in their individual pots.

The stems, though, are cut into 15-20cm (6-8in) long pieces, cutting just above a bud at the top of the cutting, and just below a bud at the base. Dip the bases in 'Keriroot' before inserting the cuttings to between one-third to half their length in individual pots of compost.

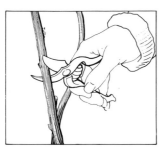

In dormant season, cut back 'parent' plant to ensure vigorous growth.

Fill small, individual pots with a good potting compost mixture and firm it well down.

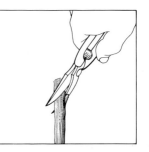

In early winter cut a number of stems with current season's growth. Slope cut above a bud.

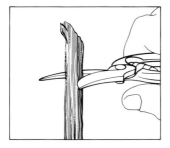

About 4cm (1½in) below the top cut, make a second, horizontally, to form base of each cutting.

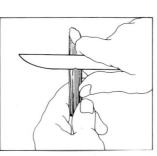

Alternatively, half-way down each stem, opposite a bud, make a shallow cut.

Carefully dip the cut areas of the stems into a rooting hormone powder to facilitate rooting.

Make a hole in the compost in each pot with a dibber and plant each cutting vertically.

Water the potted cuttings to ensure that they stay moist and then place them in greenhouse.

When the cuttings have rooted, harden them off in a cold frame. In the spring transplant them.

Layering Layering is where part of a stem is encouraged to root into the soil while it is still attached to the parent plant. The type of layering used for vines is known as serpentine layering and is carried out any time from spring to late summer.

Basically, you peg down a long, young stem into the soil in a number of places. Rooting takes place (and new plants form) wherever the stem is in contact with the soil. In this way several new plants are obtained from each stem.

The part of the stem in contact with the soil must be 'wounded' before it is pegged down. My method is to cut halfway through the stem for a length of 5cm (2in), to form a tongue and this is kept open with a small piece of wood or a small stone. Always wound at a leaf joint or node – in other words, cut through a joint. Wound the stem in a number of places along its length, and dust the tongues with 'Keriroot'.

Then peg down the stem into the soil wherever it is wounded. Peg it into a slight depression in the soil, and hold it down with a piece of galvanized wire bent to the shape of a hairpin. Cover this area with a layer of soil about 15cm (6in) deep, and keep moist at all times.

Growing on young plants Once cuttings and layers of vines have rooted they are grown on to a suitable size for planting out into their permanent positions in the garden.

LEFT Vine cuttings planted out in the open after the establishment of good roots.
BELOW Young vines beginning to make good growth.
RIGHT Healthy young grape vines raised from cuttings.

In the warmth of a propagating case vine eyes very quickly produce top growth – within a matter of a week or two, in fact. But do not think that they will also have rooted, for this takes longer.

It will probably be early spring before the cuttings have produced sufficient roots to enable you to pot them off. However, most people will be tempted to lift one or two occasionally to see how rooting is coming along. If there are not sufficient roots, re-insert the cuttings.

The rooted cuttings should be potted off into small pots – say, 9cm (3½in) diameter – using John Innes potting compost No. 1. Then they should be hardened off gradually in a cold frame, potting them on as necessary, using John Innes potting compost No. 2.

Once they are well hardened off the young vines can be stood in a sheltered spot out of doors. It will take about 12-18 months for the cuttings to reach a suitable size for planting out.

Rooted hardwood cuttings Really, the same care comments apply to these – they will quickly produce top growth but take longer to produce a good root system. Grow them on as recommended for vine eyes.

Rooted layers The layered stem should produce roots within about 12 months of pegging it down. At this stage, carefully lift it with a fork, wherever it is pegged down. If you find that it does not have much of a root system, simply replace it to allow more roots to form.

The rooted parts should be cut away from the parent plant. Make the cuts just beyond the rooted areas and discard the intermediate portions of stem.

There are two ways of growing on rooted layers. You can pot them off, choosing pots of suitable size to comfortably take the root system, and using John Innes potting compost No. 1. Then stand the pots in a sheltered place out of doors to allow the young plants to make sufficient growth for planting out.

Alternatively, the rooted layers could be planted in a spare piece of ground to grow on – in a 'nursery bed'. They can then be lifted when they are of a suitable size for planting in their final positions.

Supporting young vines Each vine will need a bamboo cane for support while it is growing. Tie in the young stem as it grows, using soft garden string or raffia.

As with most plants, grape vines have their fair share of pests and diseases, which should be kept under control to ensure optimum growth and cropping.

DISEASES

Grey mould This is a fungal disease, also known as botrytis. If it is not controlled the berries will rot, becoming covered with a greyish fungus. It is most prevalent in damp, airless conditions, and can be prevented by ensuring adequate ventilation in the greenhouse.

I would suggest also giving the vines a protective spray of Benlate + 'Activex' at flowering time, repeating this at fortnightly intervals. Stop spraying three to four weeks before harvesting.

'Huxelrebe' showing botrytis

Powdery mildew This is another common fungal disease which appears as a white powdery coating on the leaves and shoot tips. Fruits become discoloured and may shrivel.

Start spraying the vines with Benlate + 'Activex' as soon as this disease is noticed, and continue fortnightly until three or four weeks before harvesting.

Powdery mildew on grapes

PESTS

Glasshouse red spider mite Colonies of this tiny creature (which is barely visible with the naked eye) feed on the leaves, which results in very fine pale mottling. This action stunts the growth of the vines. The mites quickly build up if the atmosphere is very dry, so humidity will help to keep them under control.

However, this may not be enough to completely control the pests, so if trouble is noticed, spray the vines, at seven-day intervals with 'Sybol'. Or use 'Fumite' General Purpose Smoke, as directed by the maker.

Mealybug These soft bugs live in colonies on the young stems and shoots, sucking the sap and causing a weakening effect. They are covered in a white waxy 'meal'. Spray the dormant vines with 'Clean-Up' in the early winter. In spring and summer treat with 'Sybol', or 'Fumite' General Purpose Smoke.

Scale insects These are flat, rounded, often brownish insects and do not move around. They attach themselves to young shoots and there suck the sap. To keep scale insects under control, remove the old bark from the main stem of the vine in winter, using a knife. Use a scraping action, and be careful not to cut into the wood. Then spray the vines (in early winter) with 'Clean-Up'. To control this pest in spring and summer spray with 'Sybol'.

Vine weevil The adults are dull black beetle-like creatures with an elongated 'snout', and eat out notches in the leaf edges. The grubs or larvae can also cause a lot of damage, as they live in the soil, feeding on the roots of the vines. The grubs can be controlled by drenching the soil with a solution of 'Sybol'. The adults are more difficult to control by spraying, but try spraying them with 'Sybol'.

Whitefly These tiny white flies colonize the undersides of the vine leaves, sucking the sap and weakening the vine. Spray the vines as soon as these pests are noticed, using 'Sybol' or 'Picket'. Or ignite some 'Fumite' Whitefly Smokes, as directed by the maker.

PHYSIOLOGICAL DISORDER

Shanking This is not a disease but is caused by poor cultivation or unsuitable soil conditions. The berries shrivel and turn very sour at the ripening stage. To prevent shanking, carry out good cultivation as described earlier in this book, ensure the soil is well drained, and do not allow the vine to produce an excessively heavy crop of fruits.

OUTDOOR GRAPES

These are prone to the same pests and diseases as grapes under glass, and they are controlled in the same way. Powdery mildew particularly can be troublesome, as can grey mould, especially in a damp season.

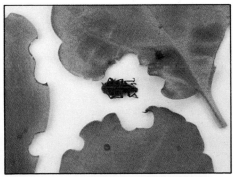

Vine weevils eat the leaf margins of vines

Vine scale insects suck the sap

INDEX AND ACKNOWLEDGEMENTS

Picture credits

A-Z Collection: 39,40
Marion Furner: 19(l,r),20,23,25(t),27,42,46(r)
Donald Smith: 10,12,14,25(b),28(b),32(l),34,36,42(r),44(t,b),47(l,r)
Harry Smith Horticultural Photographic Collection: 1,6(bl,br),7,8,9(t,b), 11,16(t,b),17,18,24,25(c),26,30(l,r),31,32(r),37(t),38(l,r),45,46(l)
Michael Warren: 4/5

Artwork by Richard Prideaux & Steve Sandilands